ROAD BUILDERS

HEAVY EQUIPMENT

David and Patricia Armentrout

The Rourke Book Co., Inc.
Vero Beach, Florida 32964

© 1995 The Rourke Book Co., Inc.

All rights reserved. No part of this book may be reproduced or utilized in any form or by any means, electronic or mechanical including photocopying, recording or by any information storage and retrieval system without permission in writing from the publisher.

PHOTO CREDITS
© Armentrout: Cover and page 4; © East Coast Studios: pages 7, 18, 21; © CASE: page 12; © Caterpillar Inc.: page 8; © John Deere: Title page, and pages 10, 17;
© Courtesy of Kubota Corporation Pages 13, 15

Library of Congress Cataloging-in-Publication Data

Armentrout, Patricia, 1960-
  Road builders / by Patricia Armentrout and David Armentrout.
    p. cm. — (Heavy Equipment)
  Includes index.
  ISBN 1-55916-136-1
  1. Road machinery—Juvenile literature. [1. Road machinery.
2. Road construction.]
I. Armentrout, David, 1962- . II. Title. III. Series: Armentrout, Patricia, 1960- Heavy Equipment.
TE223.A75 1995
                                                              95–3976
                                                                CIP
                                                                 AC

Printed in the USA

TABLE OF CONTENTS

Road Builders	5
Surveying the Land	6
Preparing the Land	9
Shaping the Road	11
Drainage Ditches	14
Graders	16
Compactors	19
Pavers	20
Finishing Touches	22
Glossary	23
Index	24

ROAD BUILDERS

At one time roads were little more than foot paths through the forest. As the population grew and new **technology** (tek NOL uh jee) was invented, roads became more and more important.

There are millions of miles of roads in the U.S. Some roads are superhighways that have as many as ten lanes. Some are small dirt roads that are hardly wide enough for two cars to pass.

A bulldozer smooths over fill dirt to prepare the surface of a new road

SURVEYING THE LAND

When a new road is to be built, many steps are taken before construction can begin. The path the road will follow depends on the shape of the land and natural features such as lakes, rivers, and mountains. Some of today's roads follow trails that were used by Native Americans for hundreds of years.

First the land is **surveyed** (SER vayed) to mark the exact path the road will follow. Work crews then begin to clear the land.

Land is surveyed before a road widening project can begin

PREPARING THE LAND

Trees and rocks are removed using bulldozers and other heavy equipment. Sometimes dynamite is used to carve tunnels through the side of a mountain. Loaders are used to scoop the extra dirt into dump trucks to be hauled away.

What happens when the path of the road is blocked by a river? A bridge must be built so that the road can cross over.

When a road is built through swamp land, truckloads of dirt and gravel are brought in so that the road can be constructed on higher ground.

Rock that has been blasted away with dynamite is hauled away in dump trucks

SHAPING THE ROAD

Land must be shaped and leveled before a road can be built. Giant machines, called scrapers, scrape the top soil off the cleared land.

Scrapers also begin the process of shaping and leveling the land. As the big machine **lumbers** (LUM berz) across the ground, a giant blade scrapes tons of earth into its oversized belly. The scraper then empties the extra dirt in low-lying areas.

This scraper collects dirt in its belly while leveling land for a new highway

A backhoe/loader empties gravel to be used in the construction of a new road

This tractor can be fitted with several different attachments making it very useful at a work site

DRAINAGE DITCHES

While the road is still in its rough stage, large machines dig **trenches** (TRENCH ez), or ditches.

Backhoes and excavators have a long arm with a big shovel attached. The shovel takes big bites of dirt and rock from the side of the road. Soon a deep ditch is dug along both sides of the road.

Large pipes are lowered into the ditches using small cranes. When the pipes are connected, they will help drain the road during heavy rains.

After digging a ditch alongside a roadway, this excavator w/ lay drainage pipes

GRADERS

When the road is ready to be smoothed over, a motor grader is driven over the unfinished road. A long, sharp blade under the grader pushes, or grades, extra dirt into big piles.

The grader operator sits in the back of the machine. From a high seat the operator can see which areas of the new road are uneven. The driver's job is to make the surface as even and flat as possible.

This grader pushes uneven layers of dirt with its shovel

COMPACTORS

Large dump trucks are loaded with gravel and stone from nearby rock **quarries** (KWOR reez). The gravel is dumped into piles on the surface of the new road.

Graders and tractors with rake attachments spread the gravel evenly over the road.

Big compactor machines with heavy rollers flatten the gravel into the roadbed. Some country roads are considered finished at this point.

A heavy roller flattens and compacts hot asphalt

PAVERS

Large roads and highways must be paved with concrete or **asphalt** (AS falt) before they are used.

Mixers pour great amounts of asphalt or concrete onto the gravel roadbed. Wide machines that move on crawler treads help spread the asphalt evenly over the road. These machines, called pavers, are so big they are brought to the work site in sections and then put together like pieces in a puzzle.

The pavers not only spread the asphalt, but they also shape the edges of the road.

Big paving machines spread fresh asphalt

FINISHING TOUCHES

The road is almost finished. Giant rollers flatten the asphalt while it is still warm. The rollers weigh thousands of pounds and compress the asphalt making it stronger and smoother.

When the asphalt is dry, stripers drive down the new road and paint traffic lines. The special trucks can paint many miles of traffic lines in a single day.

The finished road is now ready for use. Cars, trucks, motorcycles, and buses can now drive on the new road.

Glossary

asphalt (AS falt) — a dark solid plastic-like material used to pave roads

lumbers (LUM berz) — to move heavily or in a clumsy manner

quarries (KWOR reez) — holes in the earth for mining building stone, like limestone and slate

surveyed (SER vayed) — viewed, measured, and studied the shape of the land

technology (tek NOL uh jee) — useful things that result from scientific study

trenches (TRENCH ez) — ditches or hollowed out areas

INDEX

asphalt   20, 22
backhoe   14
blade   11, 16
bulldozers   9
compactors   19
concrete   20
dump trucks   9, 19
dynamite   9
excavators   14
graders   16, 19
loaders   9
lumbers   11

pavers   20
quarries   19
rollers   19, 22
scrapers   11
shovel   14
stripers   22
technology   5
tractors   19
treads   20
trenches   14
surveyed   6